DATE DUE


```
574.5    Reed, Catherine
REE      Environment
```

MESA VERDE MIDDLE SCHOOL
8375 Entreken Way
San Diego, CA 92129

SCIENCE FAIR
HOW TO DO A SUCCESSFUL PROJECT

ENVIRONMENT

BY
CATHERINE REED

SERIES CONSULTANT
DR. JOHN M. LAMMERT
Associate Professor of Biology
Gustavus Adolphus College
St. Peter, Minnesota

**ROURKE
PUBLICATIONS
INC.**
Vero Beach, FL 32964
U.S.A.

LIBRARY OF CONGRESS CATALOGUING-IN-PUBLICATION DATA

Reed, Catherine, 1948–
    Environment / by Catherine Reed
       p. cm. — (Science fair)
    Includes index
    Summary: Explains the scientific method and suggests various projects related to the environment.
    ISBN 0-86625-431-5
    1. Environmental engineering—Experiments—Juvenile literature. 2. Ecology—Experiments—Juvenile literature. 3. Human ecology—Experiments—Juvenile literature. 4. Science projects—Juvenile literature. [1. Science projects. 2. Ecology—Experiments. 3. Human ecology—Experiments. 4. Environmental engineering—Experiments. 5. Experiments.] I. Title II. Series.
TD148.R44 1993   92-12423
574 .5—dc20   CIP
                                                            AC

DESIGNED & PRODUCED BY:
MARK E. AHLSTROM
(The Bookworks)

PHOTOGRAPHY:
Cover–THE IMAGE BANK/Jean-Francois Podevin
Text–MARK E. AHLSTROM

© 1992 Rourke Publications, Inc.

All rights reserved. No part of this book may be reproduced or utilitzed in any form or by any means, electronic or mechanical, including photocopying, recording or by any information storage and retrieval system without permission in writing from the publisher.

The publisher and author accept no responsibility for any harm that may occur as a result of using information contained in this book.

TABLE OF CONTENTS

CHAPTER 1.
What Makes a Science Project "Scientific?" 4

CHAPTER 2.
Choosing the Environment As A Topic For Your Project 5
    Types of Projects 9
    The Scientific Method 12

CHAPTER 3.
Planning Your Project 16
    Gathering Information on Your Project 16
    Making Plans 18

CHAPTER 4.
Doing Your Project 29
    Safety 29

CHAPTER 5.
Presenting Your Project 39
    Final Words of Encouragement 44

SUPPLIES 45

GLOSSARY 46

INDEX 48

CHAPTER 1
What makes A Science Project "Scientific?"

When you do a good science project, you will use the **scientific method**. You will be a scientist. Scientists study the natural world and how it works. We observe the world and make measurements. We record our results with care. Then we make them into tables and graphs so we can understand them better. We ask questions about nature and plan experiments to answer our questions. When we do an experiment, we look for the effect of one thing on another, or try to find the cause of something we observe. The new discovery may give us a new way of doing things. It may also lead to new questions which we can explore with new projects. We hope that our studies of the environment will help us live in new ways. Our goal is to protect plants, animals, and the natural environment and to help human beings at the same time.

Scientists constantly teach each other about the world and the best ways to study it. When you complete a science project, you will have learned about one subject. You will be more aware of the natural environment and the ways people affect it. You will also have learned how you can study any subject in a scientific way.

When you do a science project, you are a scientist.

CHAPTER 2

Choosing the Environment as a Topic for Your Project

What will your project about the environment be? Think about lots of ideas before you decide on one. Start with something you like, one in which you are interested and that you think is important. Maybe there are some things about your environment that make you sad or angry. These could be part of your project.

There are hundreds of possibilities. The environment is a huge and fascinating subject. There are three parts to environmental science: ecology, human ecology, and environmental management.

☞ *Ecology*

Ecology is the study of how plants, animals, and **microorganisms** live together. It includes the ways that air, soil, sunshine, water, and other resources make these living things possible. When you study topics like these, you are studying ecology:

- ★ The effect of fertilizer on earthworms.
- ★ The types of birds found in the forest compared to the birds in town.
- ★ The behavior of pond snails.
- ★ Food preferences of birds at feeders.
- ★ The effect of crowding on corn plants.
- ★ Types of predatory insects in the garden.
- ★ The possibility that our climate is getting warmer.
- ★ Food webs in pond water—how do they change with time?
- ★ The effect of water temperture on goldfish respiration.
- ★ Plants and animals found in our schoolyard.
- ★ Does soil remove acid from water?
- ★ What type of soil is best for plant growth indoors?

☞ Human ecology

Human ecology is the study of how human beings live in the world and change it. Agriculture and gardening, hunting, human diseases, water, and energy supplies are all part of human ecology. Here are some examples of topics for projects in human ecology:

> ★ How much farm land has been lost to development in our county?
> ★ How can we bring back prairie plants to areas where they have been destroyed?
> ★ How much water do we use compared to that used by people in other countries?
> ★ What types of crops did our ancestors eat and how did they grow them?
> ★ How much meat do people in our city eat compared to the national average?
> ★ How many miles does our food travel to reach us? How much energy does this use?

☞ Environmental management

Environmental management is the study of how we can meet our food, energy, and waste disposal needs, while water, air, land, and wildlife remain protected. This includes developing ways to help people change their behavior and creating new inventions that maintain the environment. We always need up-to-date information to help us make decisions. Some possible topics to study in environmental management include:

> ★ What proportion of our garbage can be composted? Can worms speed up composting? How much garbage would we produce in ten years with and without composting?
> ★ How many people would ride the bus if one were available?
> ★ What will the population of our city be in ten years?
> ★ Is solar heating for homes practical in our area?
> ★ What parts of our county should be kept as nature areas? How can we manage them?
> ★ How can we start a community garden?

Be imaginative and sensible when you choose your topic. Don't pick something obvious, like, "Does fertilizer help plants grow?" Don't pick anything silly, like, "Can fish live in Kool-Aide?"

Start with something you like.

If you have *no idea* what to do, try one of the projects suggested in this book, or look at some of the books in your school library. Read the section on library research in this book. Ask your teachers, parents, or older kids to suggest ideas.

It may help to write lists of what you might like to do. You can make a list of topics you would like to study, equipment you would like to use, experiments you would like to do, things you would like to make, and field trips you would like to take. Then plan a project that will allow you to do some of these things.

Warning: Some science fairs have rules about what kinds of things are allowed to be used in projects. For example, there may be rules against using live animals or fish. It may be impossible to use electricity in your display. Check on these things before you get started.

☞ *Developing your topic*

When you have several possible topics in mind, think about which ones would be the most fun to do. Once you know what topic you like, think of as many ideas as you can and write them down. Start a new notebook, which you will use only for your project. Ask friends to help you think of ideas. Don't worry if the ideas seem strange or hard to do. Get your imagination working by thinking of many ideas. Remember, this is *your* project. Make it unique. Part of science is trying new things or seeing if old ideas work in new places.

Once you pick a topic, think of as many ideas as you can.

Types of projects

Once you have your idea, work it out and decide what to do with it. A scientist works in a logical way: First, you notice, observe, or read about something that makes you curious. Then, you start thinking of your idea as a question to be answered. Next, you find out more about the topic in various ways.

There are four different types of projects you can choose to do about the environment:

- ★ You can prepare an *exhibit* that displays a collection or shows what you have learned after reading about a topic.
- ★ You can do a *demonstration* to show how something works or is put together.
- ★ You can conduct a *survey* in which you observe activities or gather the opinions of many people.
- ★ You can do an *investigation* that uses the scientific method to investigate a problem.

Sometimes, science fair rules discourage the first three types of projects. Teachers or judges may feel that these projects are not scientific enough. However, you might be able to do an exhibit, demonstration, or survey if you add some investigation to the project. It's a good idea to have your teacher go over your project plans with you. This will help you find out if your project satisfies your local rules.

☞ *Exhibits*

One way to learn more is by making an exhibit on your topic. This is a display that people can look at. For example, these projects could be exhibits:

- ★ Freshwater life in our area.
- ★ Is our climate getting warmer?
- ★ Population growth in our county in the last 100 years.
- ★ Energy use by my family.
- ★ Insects found on different kinds of plants.
- ★ How many plants and animals can we find in one square foot?

You could also develop a plan and make an exhibit of it. For example, a plan that shows different ways your school could save paper and the amounts that could be saved might be presented as an exhibit.

An exhibit is a display that people can look at.

☞ *Demonstrations*

Demonstrations are models set up to show people how something works or what happens. For example, you could demonstrate how paper is recycled. In a demonstration, you know what is going to happen. You just want to practice some skill and display it to people.

A model can be made that shows a device that helps the environment, such as a solar cooker or a compost bin. For the science fair, design your own model. Don't make something from a kit! You can also make a model of an **ecosystem** and use it in an experiment. An aquarium can be used as a model for a lake. A terrarium can show how living things need each other to keep living. A garden could be a model for a farm or your school could be a model for all the schools in your state.

Demonstrations are models set up to show people how something works.

☞ *Investigations*

If you have a question to which you don't know the answer, you may be able to answer it with an investigation. If you have an idea *why* something happens, you can design an experiment to see if your idea is correct. You try to answer the question by using the scientific method. Questions like these could be developed into investigations:

- ★ What is the effect of pesticide spraying on beneficial insects?
- ★ What are the possibilities of using soap to kill insects on house plants?
- ★ What is the effect of water temperature on goldfish breathing rate?
- ★ What is the effect of fertilizer on pond water plants and animals?
- ★ Can worms be used to speed up composting?
- ★ Do wildflower seeds need to be frozen to sprout?
- ★ Why do dandelions near buildings bloom earlier than do dandelions in lawns?

Not all questions lead to investigations. Here are some examples:

- ★ How much garbage does our school produce?
- ★ How much energy do we use?
- ★ How many kinds of birds come to our feeder?
- ★ Which insulation works the best?

These questions are not answered with experiments because we are only observing something. They would make good exhibits, but they wouldn't be investigations.

The Scientific Method

To do an experiment, you must have some idea that one thing causes another. For example, you may have read that algae making lakes turn green is caused by fertilizer from farms. You then set up a **hypothesis**, a prediction of what you think will happen. The experiment will be the test of the hypothesis. For example, you may think that algae grow in lakes because too much fertilizer comes into the water. This is a hypothesis.

To do the experiment, you need an **experimental design**. This is a plan to test the hypothesis, that is, to see if the hypothesis is correct. In this case, your models of the lakes could be aquariums full of clear pond water. One group of aquariums will be treated with fertilizer. This is the **experimental group**. You add different amounts of fertilizer to separate aquariums. The amount of fertilizer is called the **independent variable** (something you change on purpose). Several aquariums are not treated. This is the **control group**. The control and the experimental groups are exactly the same when the experiment starts. Any differences that develop during the experiment must be due to the independent variable, the condition that you change.

You must have more than one experimental aquarium and more than one control aquarium. This is true for any experiment. Just one test is not enough to tell you whether any differences you see after adding the fertilizer were due to the fertilizer.

You will make **observations** of what is happening. For example, whether the water turns green or stays clear will be your observation here. If a lot of algae are present, the water will turn green. In this case, the color of the water (or the growth of algae) is called the **dependent variable** (the one that changes in response to the independent variable).

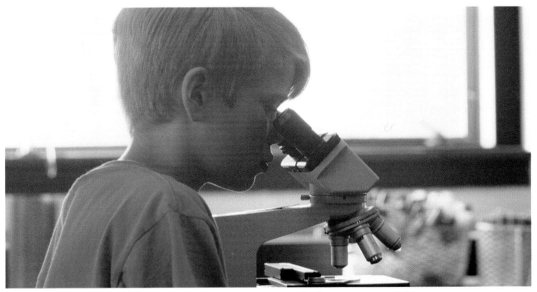

Microscopes open up a whole new world!

The experiment will give you **results** in the form of measurements or observations. Another word for results is **data**. You will need to **analyze** the data to find out what they mean. Then you will have a **conclusion**. Suppose that all of the control aquariums remain clear and all of your experimental aquariums turn green. You might conclude that fertilizer caused the green color. If you then look through a microscope and see algae in the green water, but no algae in the clear water, you can then also conclude that algae growing in the water cause the green color.

The conclusion always comes from the results. It is an answer to the question with which you started. Your conclusion might agree with your hypothesis. If so, you can say that the experiment supported the hypothesis. Sometimes your results will not be what you expected—still, the conclusion must agree with the results.

Using all these steps to test a hypothesis is called the scientific method.

Steps Followed in the Scientific Method

✔ Make an observation.

✔ State the problem: What do you want to find out?

✔ What is already known about the observation?

✔ Develop a hypothesis: What do you think is a reasonable explanation for the observation?

✔ Design an experiment that will provide answers: What materials will be needed and how will they be used?

✔ Record data or observations: What happened during the experiment?

✔ Analyze the results.

✔ Draw a conclusion: What did you learn? Did you find out what you wanted to know? Did your data support your hypothesis? What do your results mean?

Maybe the experiment will tell you your hypothesis was correct. Maybe the experiment will show that your hypothesis was wrong. For example, the fertilized aquariums might stay clear and the unfertilized ones might turn green. Maybe the experiment didn't work at all and you won't be able to tell anything from your results. Whatever happens, you will have learned something new.

Probably new questions will come up as you do the experiment or when you are analyzing the data. Let these lead you to another project.

Environmental science projects may also have **implications** for actions. These are suggestions of what to do to solve a problem. If overuse of fertilizer can cause algae to grow quickly and turn a lake green, fertilizer needs to be prevented from washing into lakes. Here is another example. Let's say your project shows that your county will run out of landfill (dump) space in a few years. The implication is that everyone will have to reduce the amount of garbage they throw away or find something else to do with it. Maybe your question asked if a bottle deposit law would reduce roadside litter. You find out that 40% of litter along streets and highways is bottles. You also find in a survey that most people agreed that they were willing to return the bottles for a deposit. The implication then is that a bottle deposit law would reduce litter.

CHAPTER 3
Planning Your Project

Gathering Information on Your Project

☞ Library Research

Start by writing down what you know about your topic. What else do you need to know? Reading books and magazines found in a library will help you learn what other people have discovered about your topic and about ways to study it.

Ask the librarians to help you find information. They will be happy to help you to use the card or online catalog to find books on your subject. They can help you use the "Reader's Guide to Periodical Literature" which lists articles found in magazines. This reference book arranges articles by subject.

Explore the library thoroughly. Use several sources, not just one book or encyclopedia. Don't believe everything you read. Remember that some authors may have special reasons for their opinions. For example, if the author works for a plastic company, the book may favor plastic recycling.

Use several sources, not just one book or encyclopedia.

Another author might think that all plastics should be banned. If something an author writes seems wrong or questionable, take some time to think about it. You might think of an investigation to see if the author is right or wrong. Some things that work in one part of the country may not be practical where you live.

Make notes of what you read in your special notebook. For each book you read use a separate page. Write down the title, author, date, publisher, and city of publication at the top of the page. For magazine articles, record the name and date of the magazine, the author (if any), and the pages where the article appeared. You will need this information when your make the **bibliography** in your written report.

When you take notes, write down the main points. You don't need complete sentences. Concentrate on the information that will help you answer your question. Use your own words. It is not honest to use the sentences that an author writes when you finally present your project. This is called **plagiarism**.

☞ *People who can help you*

Most scientists work in groups so they can help each other and get more work done. You also may be able to do this. Check the science fair rules to see if you are allowed to work with classmates. For big projects, like starting a school garden, a nature center, or recycling program, you will need help from other people. Your teacher, a high school science teacher, parents, and older kids who have done science projects can all help.

For information on ecology, check with museums, colleges, agricultural extension services, your state Department of Natural Resources, nature study clubs, and conservation groups. For environmental issues, the Sierra Club, National Wildlife Federation, Nature Conservancy, and Audubon Society may have information. Get their addresses and telephone numbers from the library.

People in state and local government will have information on local environmental issues and local history. Recycling centers and food coops may be able to help. People in some businesses, such as solar heating and biological pest control, may be able to help with their specialties.

Don't be shy about asking adults for information and help. Everyone agrees that it is very important for young people to be informed about the environment. Usually, if one person can't help you, he or she will try to think of someone else who can. Send everyone who helps you a copy of your report when you finish.

If people disagree with each other, try to figure out why. Maybe each person has some good points and maybe each is partly wrong. Try to get more information without arguing.

Making Plans

Now that you have decided what your project will be, you need to plan exactly what to do. Write down everything in your notebook.

First you have to decide what question the project will answer. If you are doing an investigation, state your hypothesis. Then decide what data you will need to collect to answer your question and how you will get it. The way you collect the data is called **methods**. Methods includes everything you do to get your information. Determine what materials you will need and find out where you will get them. Then decide how you will collect your data as results. Your results will help you arrive at a conclusion. In your conclusion you will analyze your data and explain what they mean. Think about what you might do the next time if things did not turn out the way you expected. Finally, think about how you will display the data so other people can understand them easily. It may help to make a **flow chart**. This shows each step in the project, in the right order:

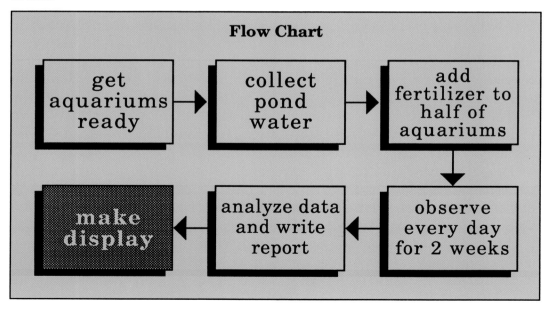

It may also help to set up a timeline or schedule. Make a definite deadline (or due date) for each part of the project.

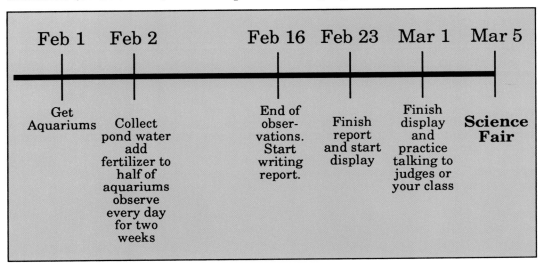

The following sections will help you make your plans.

☞ *The question and data to answer it*

The question is probably the most important part of your science project. Your question must be something you can try to answer with the materials and time that you have. If you are doing an investigation, you must predict what the answer will be, based on what you already know. This is your hypothesis, which will be tested by your experiment.

Data is any kind of information you collect to answer a question. They can be either observations or measurements. If you are doing an investigation, the data will either support or go against your hypothesis.

Suppose your question asks "How much food do students in my school waste at lunch each day?" Your data would be pounds or kilograms of food per student per day. If your question asks "How many people would ride the bus to the mall if there were one?", the data would be the percentage of people you surveyed who said they would do this.

What if your question is "Could we use solar collectors to heat our house?" Data would include the types of collectors available, how well each works in your climate, and how much it costs to use a collector compared to your present heating bills. For the question "Does soil remove acid from acid rain?", data would be how acid the water is before and after it has passed through different soils.

You use words, measurements recorded in tables or graphs, or photographs to record data. Be sure to keep all your data in your lab notebook. When you write something down include the date. You will learn more about recording in Chapter 4.

The kind of question you have asked determines the methods you need to use to get your data. Some questions can be answered by library research, some by asking people, some by observation, and some by doing an investigation.

Don't be shy about asking adults for information and help.

☞ *Planning library research*

Decide exactly what information you need to answer your question. For example, if you want to know how much farmland has been lost in your county over the past 20 years, you will have to find out how much farmland was present 20 years ago and how much is present now. Probably you will need an old reference book and a new one. Suppose your question asks "Is our climate getting warmer?" You must decide if you need average temperatures for your whole state or just a part of it. You will also need to decide if you want averages for one year at a time or for ten years at a time.

☞ *Planning ways to get information from people*

If you are doing a survey, you might have to take a **poll**. This means that you ask people what they think about an issue or idea. You want their opinions, not facts. Decide exactly what your questions will be, who you will ask, and how to write down what they say. It's easiest if you write down the questions and give a choice of possible answers. This is a form called a **questionnaire**. Make many copies, like your teacher does for a school test. Write down or circle each person's answers on a different copy. Remove people's names when you make your report. Here is an example of a questionnaire:

Questionnaire: Riding the Bus to the Mall

Name _____ Address _____ Date _____

Distance you live from the mall _____

How do you get to the mall now? (circle one):   walk   bike   drive

If the city started a bus to the mall would you take it? (circle one):

       at least once a week   once a month or less

       once every two weeks   never

When you have collected as many answers as you have time for, put the results in a table so you can add them up. The blanks in the table below are for writing in the number of people in each group.

Table 1. People Who Said They Would Ride the Bus

Once a week or more	_____
Once every two weeks	_____
Once a month or less	_____
Total who would ride	_____

Maybe their answers depend on how far they live from the mall. You can make a table to check this.

Table 2. People Who Said They Would Ride the Bus

	Once a week	Every 2 weeks	Once a month	Total Riders
Distance from Mall:				
Less than 1 mile	_____	_____	_____	_____
1 to 5 miles	_____	_____	_____	_____
5 to 10 miles	_____	_____	_____	_____
Total	_____	_____	_____	_____

Add up a total for each group. Then you can say something like, "Seven out of 10 people who live between 1 and 5 miles from the mall said they would take the bus once a week, but only 1 out of 8 people who live 5 to 10 miles away said they would take it that often."

You can get data by doing a survey.

If you ask questions that don't have short answers, record them in your notebook. For example, if you call the mayor's office to ask about recycling in your community, write down the date, name of the person with whom you talked, questions you asked, and notes on what this person said. It may help to write down your list of questions before you call so you don't forget anything.

Writing letters is another way to get information. Keep your letters short and as specific as possible. State exactly what information you need. Say why you need the information and include the date you need it. If you don't receive an answer within two or three weeks, make a phone call.

☞ *Planning observations*

Observations are needed for all investigations. They also might be used in some surveys. Decide what kind of and how many observations you will need to make. You should make as many observations as you can—at least three or four. Decide what the units of measurements will be. Make sure your observations are related to your purpose or hypothesis. If your observations involve measurement, use the metric system.

Make a data table in your notebook to record your observations. Here are some examples of data tables with spaces to write in the observations. Leave space on your tables to write down anything that was unexpected.

Data Table for Washington School Lunchroom Garbage Survey

Date _____
Class _____

Pounds of Garbage:

Paper	_____
Plastic	_____
Food	_____
Other	_____
Total	_____

Data Table for Flowers Bees Like Best
(Bees Observed on 100 Flowers of Each Kind)

Date:	Bees on Goldenrod	Bees on Aster	Bees on Red Clover	Total Bees Seen This Day
Total for Each Flower:				

You could make copies of the above table to use for several days of collecting data. This will tell you a lot more than only one day of making observations. There is space at the bottom for totals. You could also calculate the percentage of bees found on each kind of flower, or the average number of bees on one kind.

On the following page is an example of a table that could be used to record observations made during an investigation. Each trial is one time you did the experiment. You can take an average of the trials for each temperature. The heading will remind you to do each trial the same way.

The Effect of Temperature on Breathing in Goldfish Number of Times The Goldfish Moved Its Gills During a 30-Second Observation					
Water Temperature in Degrees C	Trial 1	Trial 2	Trial 3	Trial 4	Average for each temperature
18					
24					
28					
32					
Average for each trial					

☞ *Planning experiments*

First you need a hypothesis and an experimental design. You must decide on experimental and control groups or areas, how to get your data, and how to make your analysis. Look at Chapter 2 for more information on this. You need an experiment if you have a new question and you don't know the outcome.

Here is an example. You have an idea that bees visit flowers to get sugar. You make a hypothesis: bees visit flowers with the most sugar. Then you decide on an experimental design. Your plan will be to add sugar water to some flowers. The flowers that get the sugar water will be the experimental group. You will need another set of flowers that do not receive any sugar water. This is the control group. You will observe how many bees come to the flowers with sugar water and how many bees come to the flowers without sugar water. This information is your data. You have to decide if there is a difference in the number of bees visiting the two groups. This is your analysis.

Be careful when working around bees!

Here's another example. Suppose you find a lot of sowbugs in your basement. You think they are there because it is damp there. You make a hypothesis: sowbugs prefer damp areas. To test your hypothesis, you divide a box into two parts—one damp and one dry. You release sowbugs in the middle. After 12 hours, you count how many go to the damp side and how many go to the dry side. You repeat this procedure several more times. The number of sowbugs on each side is your data. You then calculate an average of how many sowbugs are found on each side—this is the analysis.

☞ *Finding materials*

Make a list of the materials you need. You might be able to borrow special materials from a high school or nearby college. You could order them from a science supply company. Some addresses are listed at the end of this book. Your teacher may have to place the order for you. Science supply companies sell live plants and animals, as well as equipment and chemicals. Check local hardware stores before you order special equipment. It may take at least two weeks for your order to arrive by mail.

☞ *Planning you bibliography*

When you do your written report, you will have to include a bibliography (see Chapter 5). This is the way most scientists list books and magazines when they do research:

For a magazine:
   Barinaga, Marcia. "The secret of saltiness." Science 254:664-665 (1991).

For a book:
   Hodges, Laurent. Environmental pollution. New York:Holt, Rinehart and Winston, 1977.

CHAPTER 4
Doing Your Project

Safety

"Safety first!" is the rule when working on a science project. Here are some tips to keep you from possibly being injured.

A responsible adult should check your project plan for possible hazards. If you are going out in the field to observe or collect, get your parents' agreement first. Tell them where you are going, what you plan to do, and when you will return. Let them decide whether it is safe for you to go alone, with a friend, or with an adult. If you go from door to door asking people questions, be sure to have an adult with you.

Have a place of your own to work on your project. It might be a table in your classroom that your teacher assigns to you. The basement, a garage, a storage shed, a porch, or your room at home is suitable. Keep your work area neat. As you do your project, be sure to clean up. Put away things that you no longer need.

If any chemicals are used in your project, use them *only* with an adult to supervise you. Wear protective glasses, rubber gloves, and a rubber apron. Wash your hands each time you finish using the chemicals. Keep your hands away from your mouth while you work. All chemicals should be clearly labelled and stored in a suitable container.

If your project requires that you work outdoors in the woods or a field, watch out for poisonous plants, like poison ivy. If the leaves of these plants touch your skin, it may get red and very itchy. To reduce the chances of getting these plant oils on your body, wear long-sleeved shirts and pants with legs. If ticks are present, include insect repellent. It may be a good idea to have an adult along when you visit a forest to work.

If you choose to study algae collected from a lake, it may be a good idea to have someone come along who knows about water safety rules. Don't go wading in waters where the bottom is not familiar to you.

Any electrical equipment should use batteries to power it. Electricity from a wall socket is too dangerous to use. Even though the current available from batteries is not strong, wires coming from them can become hot and thus possibly cause a burn.

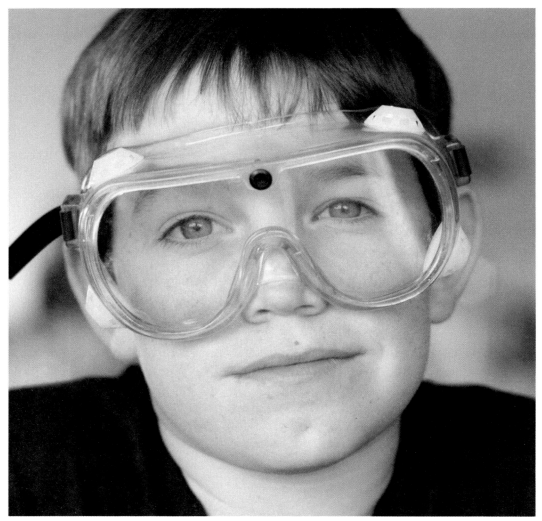

It's better to be safe than sorry!

☞ *Working on your project*

Once you have completed your plan and have your materials together, you can really get going.

Here are some suggestions that will be useful. Find a place to work where your materials won't be disturbed. Explain everything you are doing to your family. Work on your project for at least a few minutes every day—make a definite time for doing this.

Follow your plan. If something isn't working, try to figure out why before you change your plan. Write down the reasons for any change in your notebook. Record everything. Write down exactly what your methods were so another person could try your experiment. Record your observations as carefully as possible.

Work slowly and believe your results. Even if it seems to be harder than you expected, complete the work. You will be glad when it is done. Science has fun parts and boring parts, just like anything else. It's not easy to make new discoveries.

☞ *Recording your methods*

Be exact. For example, don't just say, "I added vinegar to water and poured it on plants." Instead, list every step:

1. I mixed 2 teaspoons of white vinegar with 1 cup of tapwater.
2. I poured 1/4 cup of the solution on each plant every other day.
3. The plants were 4-week-old bean plants growing in 4-inch pots of soil. There were 10 plants in the experimental group and 10 plants in the control group.
4. The control group got 1/4 cup of plain tap water every other day.
5. I measured all the plants every day for 2 weeks.

Write everything down in your notebook. Be sure to give exact measurements.

☞ *Recording your data*

How you record your data will depend on the type of data you have. Write down your observations in your notebook. Before you begin to collect data, think about what your observations will be so you won't forget anything. If possible, make out a data sheet. In chapter 3, you will find examples of data sheets.

Numbers and measurements can be recorded in tables. This will make it easy to make graphs and figure out averages later. Be sure to title the table and list the units of measurement and the date, as shown. You may want to graph the data as you record it. This makes it easier to understand what is happening.

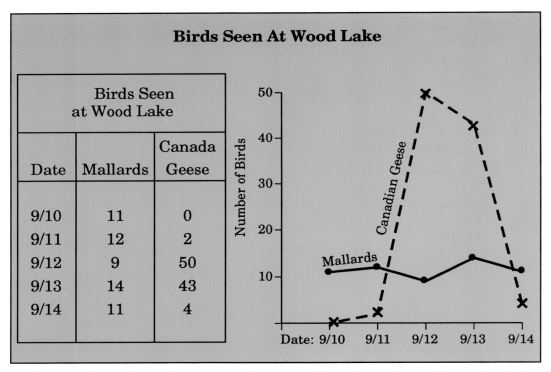

The graph on the right was made using data from the table on the left.

☞ Making graphs

We use graphs to understand our results, to show results to other people, and to get more information from our data.

Use a **bar graph** to compare groups. The length of the bar shows the size of each group. Be sure to show your units of measurement on both **axes** (lines). Give your graph a title.

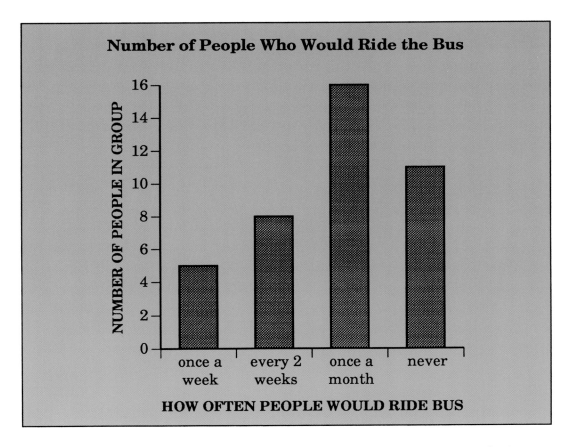

Use a **line graph** when you show observations over a period of time, or under changing conditions. In scientific terms we use a line graph when there is a change in the dependent variable (the one you measure) when the independent variable is changed. Make a dot (point) for each observation, then connect the dots with a line. The line gives an idea of what is happening between the times we made observations, although we can't always be sure of this. Be sure to give units of measurement. Title your graph.

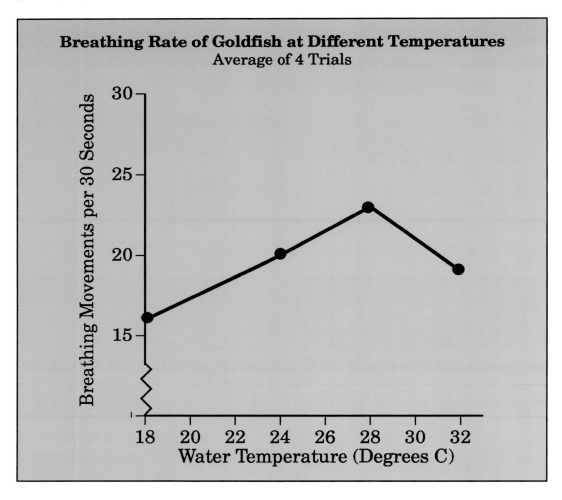

Use a **pie graph** to show parts of a whole. The data you are graphing must add up to 100 percent. To make a pie graph, you need a circle (the pie) and a protractor to measure the angles. The angle is the size of the pie wedge. Remember that a circle has 360 degrees. If you cut a circle into 100 little wedges, each one would have an angle of 3.6 degrees. Convert your results to percentages. Multiply each percentage times 3.6 to give the degrees. For example, 20 percent would equal 72 degrees (20 x 3.6). When you have the data ready, mark the angles on your pie. Start at the top.

Continue until all the angles are done. This should fill the pie. Be sure to label each wedge and give your graph a title.

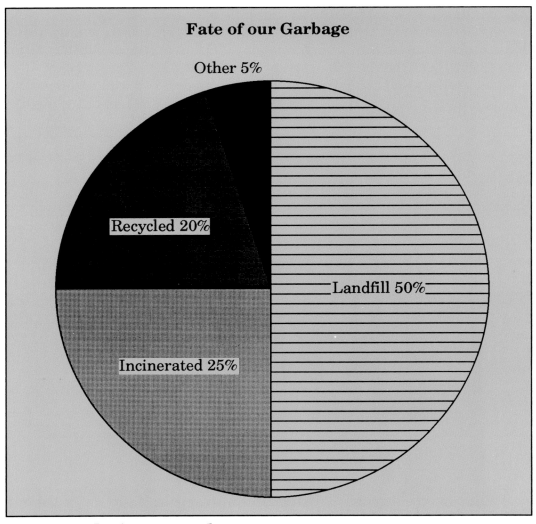

☞ *Analyzing your data*

Once you have collected your data, analyze it. Go back to the question with which you started. Can you answer it using your data? Look at your graphs or tables. Can you explain them in words? Does this explanation answer your question?

If you have information about different groups, look at each group separately and try to compare them. If you have measurements, you may want to take an average for each group. You can take averages only on groups that have been treated in the same way.

Look carefully at your results. In the example below, you wouldn't know much if you looked at just one day. The averages are different for each grade, even though the results are not the same for each day. The averages tell you more than each day's results do. You can also see that on some days everyone wasted more food than they did on other days.

Food Wasted By Different Grades Pounds of Food Wasted					
Grade	9/12	Date 9/13	9/14	Total for each grade	Average for each grade
1	1.0	1.2	2.0	4.2	1.40
2	1.5	1.0	2.1	4.6	1.53
3	1.3	1.3	1.8	4.4	1.46
4	1.4	1.3	1.6	4.3	1.43
Daily Total Daily Average	5.2 1.3	4.8 1.2	7.5 1.8		

Data tables help you analyze a problem

What do your graphs, tables, and averages tell you? Is there a difference among groups, no difference, or can't you tell? It may be hard to decide. A very small difference may be due to chance.

Even if you can't answer your original question, you can still report on what you have learned. For example, suppose you tried to use earthworms to make compost, but all your worms died. You still know something about composting and about earthworms, so put that in your report. Suppose you tried to mark bees with paint to see how long it took for them to come back to your flowers. The marked bees never returned. You now know that bees fly a long way when they are disturbed.

☞ *Making conclusions*

What do you know from your study? If the results did not support your hypothesis, was the hypothesis wrong? Was there an error in the experiment? Remember, you are trying to discover what is going on, not trying to prove your point.

Your conclusion must come from your data, not from books or from what someone told you. For example, let's say your hypothesis was that bees will visit flowers with more sugar. You added sugar water to some flowers, and you observed more bees on the flowers with added sugar. Your conclusion would then be that sugar does attract bees to visit flowers. What if you observed that there was no difference between flowers with sugar and flowers without sugar? You would conclude that added sugar does <u>not</u> attract bees. If there was only a very small difference between groups, you might not be able to make a conclusion.

If the result is not what you expected, try to figure out why. Sometimes, more observations will lead to different conclusions.

☞ *Implications*

Your conclusions might suggest that it would be good to take some kind of action. These would be the implications of your data. Suppose you found that many people wanted to take a bus to the mall. The implications might be that the city should study whether it could afford to have buses. If your study found a lot of food waste in school, the implications could be that the cooks should consider giving smaller servings. If you found that butterflies visit certain flowers, the implications would be that these flowers should be planted to help conserve butterflies. Put the implications in your report.

When you have finished analyzing your data, give yourself a pat on the back for all your work! Usually, doing a project will leave you with lots of questions that could be answered by another project. Make some notes about these, just to be ready for next year.

Give yourself a pat on the back for your hard work.

CHAPTER 5
Presenting Your Project

☞ ***Writing a report on your project***

Your report should be written so carefully that anyone who reads it will know exactly what you did and how you did it. The reader will also know why you made your conclusions, where you got your information, and who helped you.

When you write a scientific report, always state where you got your information. Include the author's name, if possible. For example, you could write "Sand pines need fire in order to drop their seeds (Feinsinger and Minno)." Or you could say something like, "The information on pines comes from <u>Handbook to Schoolyard Plants and Animals</u> by Feinsinger and Minno." Then, if anyone questions your information, they can check the book, which will be listed in your bibliography.

If you use someone's exact words, put them in quotation marks. If someone gave you an idea or explained how to do something, give them credit. For example, "Sarah Vest suggested that I look for insects at Afton State Park."

Type or write neatly. Put the report in a binder.

| The Effects of Light Color on Growth of Pea Plants

Emily Johnson
Grade 6
North Elementary School
Petertown, Minnesota | **Table of Contents**

Introduction
   Statement of Purpose 3
   Research 4
   Hypothesis/Predictions 9
Methods/Experimental Design 10
Data and Analysis 15
Conclusions 20
Implications 21
Bibliography 22
Acknowledgments .... 23 |

This is a sample title page and table of contents for your written report.

The report should have these sections:

- ★ Table of Contents.
- ★ Title page. List your project title, your name, grade, teacher, and school. If the report is more than 5 pages, put in a table of contents.
- ★ Introduction. This will have information on the project, why you did it (purpose), and your hypothesis. Other information in the introduction includes what is already known about the subject, the questions the project was planned to answer, and your prediction of what the results would be.
- ★ Methods and Materials. This tells exactly what you used and did to answer your questions.
- ★ Results. This section tells what you observed and how you analyzed your observations. Put all your records, data sheets, tables, and graphs in this section.
- ★ Conclusions. Say what your data mean. Note whether they agreed with your hypothesis.
- ★ Implications. Suggest any action that might be taken based on your conclusions. You may go beyond your conclusions with new ideas. Maybe you found that fewer birds are found in your county now than there were 20 years ago. You might think that this decrease in birds may be due to more people living in the area now.
- ★ Bibliography. List all the books and other sources of information you used. Give the author, title, date, publisher, and city of publication for books. Arrange the titles in alphabetical order by the author's last name. Include videos and television shows you may have watched and people with whom you talked.
- ★ Acknowledgements. Thank anyone who helped you with the project, and say what each person did. For example, "I thank Mrs. Holland for helping me find the spiders, and Mr. Huang for helping me make the circuit boards."

☞ *Making a display*

Most displays will have three panels and sit on a table. You can use pegboard, plywood, strong cardboard, or foamboard to make your display. Write or draw items on separate sheets of paper, then tape or glue them to the display. You might be able to buy display boards at an office supply store.

Include the real things you used, if possible. Check the rules of your science fair to see what is allowed. Your notebook should be available. Photographs and drawings add interest. Show data in charts or graphs. Make them big and easy to read. Use color. Make your lettering as neat as possible.

Every part of the project must be presented, just as in your written report. Don't give all the details, however—just the main points. Feature your question and your results.

If you are only doing an exhibit or demonstration, be sure it looks good and that everything is working. You should have a sign with the title and purpose of the project on it. Have enough materials so you can do your demonstration several times for visitors. Let visitors try it themselves, if possible.

If your project includes a poll or questionnaire, ask visitors the same questions. Add their answers to your graph or table when you have collected enough. Compare their answers to those you collected in other ways.

All displays should show your name and school.

Displays have to be neat!

☞ *Talking to judges or your class*

You will have to talk to judges and maybe to your class. The most important thing to remember is this—focus on the main points. Don't give a lot of details. Start with why you chose the project. Then tell how you did it and what the results were. Show your materials and point to your pictures or graphs as you go along.

Take your time. Be ready to answer questions. If you don't know the answer to a question, make an educated guess if you can. If you have no idea, just say, "I don't know." You might suggest a way of finding the answer, just as a teacher does in class.

Practice giving your talk to family and friends. It might help to write some notes on little cards to use during the talk.

Usually you will see between two and five judges. The judge will usually ask you your name, and then leave you free to explain your project. In four or five minutes, you have to:

- ★ Give your name and the title of your project.
- ★ Tell how you got the idea for the project and what your purpose was.
- ★ State what your hypothesis or prediction was.
- ★ Tell how you did the project and anything unusual that came up.
- ★ Report what the results and conclusions were.
- ★ Tell whether or not you expected your results and why certain things happened.
- ★ Tell how the project could have been made better and what you learned.
- ★ Tell what you would like to build on if you were to do another investigation about the same topic.
- ★ Thank the judge when she or he leaves.

When you are talking to judges, don't chew gum or pick your teeth. Be dressed nicely and try to have good posture. Know what you want to say before you say it. Look the judges in the eye. Give them the feeling that you don't have a doubt in your mind. Answer any questions honestly.

Learn from experience. Hopefully, your presentation will improve each time you talk with a judge. Put more emphasis on what the previous judges had questions about and any parts of your project that might have seemed unclear. Keeping the interest of the judges is important.

Emphasize the main points.

☞ How projects are judged

Usually the judges will come around one at a time, giving your project points on a judging sheet. There are usually points for each of the following areas. The most points are usually given for creativity and use of the scientific method.

> ★ Creativity—How unique or original is your idea?
> ★ Use of the scientific method. If your project is an investigation, do your results support your conclusions? Were you careful about using methods and controls?
> ★ Understanding—How complete or thorough is the project? How much background material do you know? Do you understand what you have learned?
> ★ Clarity—How well did you present your project? Can people understand your display?
> ★ Quality—Does the display look good? Are there mistakes in your written materials? How much of the work did you do yourself (parents can help, but you should do most of the work)?

After the judges have seen all the projects, they will meet in groups. They will add up the points and decide which projects are the best. At big science fairs there may be a long wait before they come around with ribbons or other prizes. There may be an awards ceremony. At local science fairs winners usually receive ribbons. At state science fairs the prizes may be books, pins, or even money. If you continue to do science fair projects when you go to senior high school, you can compete for big prizes like trips and scholarships.

After the fair you'll be tired and happy. While things are still fresh in your mind, go over what you have learned. Would you do this kind of project again? What were the good and bad points? Could any of your results be put into action right now?

What a great feeling!

Final Words of Encouragement

Enjoy science! You have learned many new skills by doing a science fair project. You can now ask better questions about the natural world and develop ways to answer them. You can find information and plan experiments. You can analyze your results and decide what to do about them. You can take action to improve our environment based on your new knowledge.

Good luck with your project. See you at the science fair!

WHERE YOU CAN BUY SUPPLIES FOR EXPERIMENTS ABOUT THE ENVIRONMENT

Carolina Biological Supply Co.
2700 York Road
Burlington, NC 27215
1-800-334-5551 (East of the Rockies)
1-800-547-1733 (Rockies and West)
1-800-632-1231 (North Carolina)

Connecticut Valley Biological Supply Co., Inc.
P.O. Box 326
Southampton, MA 01073
1-800-628-7748 (U.S.)
1-800-282-7757 (Mass.)

EMD
A Division of Fisher Scientific
4901 W. LeMoyne Street
1-800-621-4769
1-312-378-7770

Nasco
P.O. Box 901
Fort Atkinson, WI 53538-0901
1-800-558-9595

Science Kit & Boreal Laboratories
777 East Park Drive
Tonawanda, NY 14150-6784
OR
P. O. Box 2726
Santa Fe Springs, CA 90670-4490

Ward's
P.O. Box 92912
Rochester, NY 14692-9012
1-800-962-2660

GLOSSARY

analyze - to find out what something means.

axes - the lines on the side and bottom of a graph.

bar graph - a graph that uses columns to compare different values obtained for experimental groups. The height of each bar is proportional to the value.

bibliography - a list of books and magazines used to get information.

conclusions - what you interpret the results of an experiment to mean.

control group - a group in an experiment in which as many variables as possible are kept constant because they could affect the outcome of the experiment.

data - the observations and measurements that you make in an experiment.

dependent variable - the factor or condition that changes as a result of the presence of, or a deliberate change you make in, the independent variable.

ecology - the study of how plants, animals, and microorganisms live together.

ecosystem - all of the things that affect life in an area.

environmental management - the study of how people can live without destroying the environment.

experimental design - the plans you make so you can do an experiment. The design includes what you will use and how you intend to use them.

experimental group - a group in which all variables are the same as those in the control group *except* for the factor that you are following in your experiment.

flow chart - a list that writes out a shortened version of the steps you want to follow in doing your experiment. As you complete each step, you should check it off the list.

human ecology - the study of how human beings live in the world and change it.

hypothesis - a statement that gives a possible answer to a question, sometimes called an "educated guess." To see if it is true or not, a hypothesis is tested by doing an experiment.

implications - ideas to help solve a problem.

independent variable - the factor or condition that you want to study. In an experiment, you intentionally change this factor.

line graph - a graph that uses a line to see if the dependent variable changes as the independent variable is changed.

methods - the ways you collect data to get results.

microorganisms - very small living things, like microbes, that can't be seen without using a microscope.

observations - the things you observe during an experiment.

pie graph - a graph that uses slices of a circle to show parts of a whole; also called a circle graph.

plagiarism - copying word-for-word what someone else has written and not giving credit to that person.

poll - gathering opinions by asking questions.

questionnaire - a list of questions used when doing a poll.

results - what you measure or observe as an experiment is carried out.

scientific method - a systematic strategy scientists use to discover answers to questions about the world. It includes making a hypothesis, testing the hypothesis with experiments, collecting and analyzing the results, and arriving at a conclusion.

INDEX

Average, calculating 28, 36

Bibliography, making 17, 40

Conclusions, coming to 14, 37

Data 14, 19-20, 32, 35-37
Data tables 22, 24-26, 32, 36
Displays
    construction 41
    organizing 41

Ecology 5-6, 17
Environmental management 6
Experimental design 13, 26
    variables 13
    controls 13
    groups 13

Flow chart 18

Graphs, types of 33-35
    bar graph 33
    line graph 33-34
    pie graph 34-35

Hypothesis 12-13, 15, 19, 26-27

Implications 15, 37

Judging 42-44
    talking to judges 42
    scoring of project 43-44

Library, information in 16-17, 21
    card or on-line catalog 16
    Reader's Guide to Periodical Literature 16

Materials 28, 45
Models 11

Notebook 8, 17
Notes, taking 17, 24

Observations 13, 24-25

Polls, taking 21
Procedure 26-28
Projects, types of 9-12
    demonstrations 10-11
    exhibits 9-10
    investigations 12
Purpose of investigation 18-19

Questionaires 21

Record keeping 8, 17, 24, 31-32
Report, written 39-40
Resource people 17-18, 24
Results, analyzing 14, 22-23, 35-37

Safety 29-30
Scientific method 4, 12-15

Time schedule 19